STARSHIP Family Journey

Star Magic Notebook

Write on the stars, planets, galaxies, and nebulae!

by Laura S. Woodmansee
Chief Visionary, StarshipFamilyJourney.com

This notebook belongs to:

Download your FREE digital wallpaper to match the images in this notebook:

www.StarshipFamilyJourney.com

Get onboard our e-mail list and get more cosmic goodies!

The SFJ Star Magic Notebook: ISBN-13: 978-0-9863172-7-9
1st printing: 12/2017 by Stella Creo inc.

Stars are magical. Their light beams out stories and secrets of ancient times and distant places in our galaxy and beyond.
Look into the night sky for the first star to twinkle 'hello' to you.
Wink back at your new star friend!
And then, get cozy, be patient, and watch and listen for the secrets your star wants to tell you. What does it say?
Write or draw about it. Create your own magical star tale.
-- Laura S. Woodmansee

Second star to the right and straight on 'til morning.
-- J.M. Barrie, Peter Pan

NGC 6946
Spiral Galaxy
Image Credit: NASA/HST

Other intelligent lifeforms will differ greatly in appearance.
They may resemble the creature E.T. or startle us with their beauty.
But life itself is common, I'm certain.
-- Frank Drake

SMALL MAGELLANIC CLOUD
Image Credit: NASA/HST

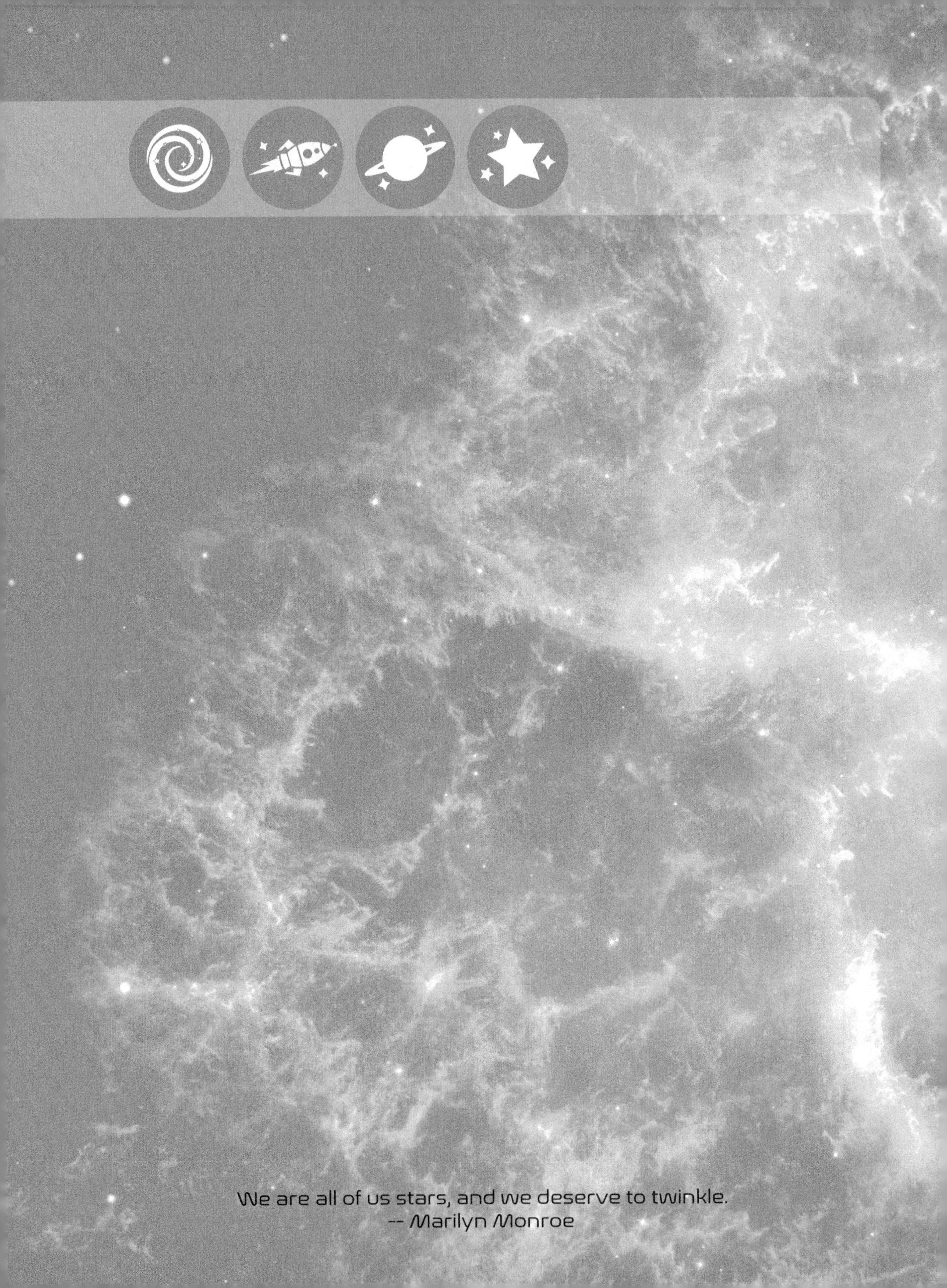

We are all of us stars, and we deserve to twinkle.
-- Marilyn Monroe

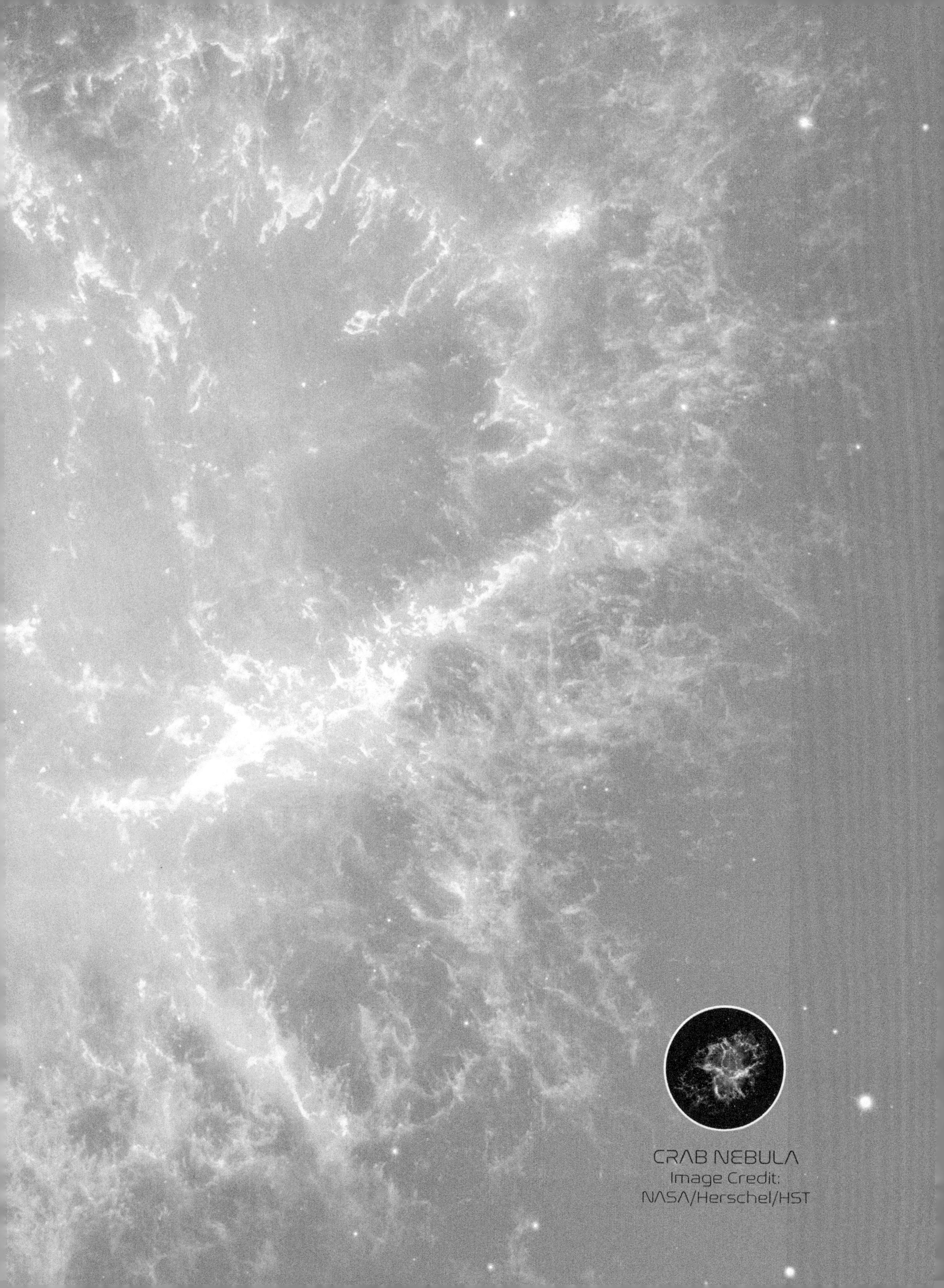

CRAB NEBULA
Image Credit:
NASA/Herschel/HST

We came all this way to explore the Moon,
and the most important thing is that we discovered the Earth.
-- Apollo Astronaut William Anders

WESTERN GLOBE
EARTH
Image Credit: NASA

We are the universe made manifest,
trying to figure itself out.
-- Delenn, Babylon 5

LOBSTER NEBULA
NGC 6357
Image Credit: NASA/HST

We are part of this universe; we are in this universe. But perhaps more important than both of those facts, is that the universe is in us.
-- Neil deGrasse Tyson

HORSEHEAD NEBULA
Image Credit: NASA /
Hubble Space Telescope

For my part I know nothing with any certainty,
but the sight of the stars makes me dream.
-- Vincent Van Gogh

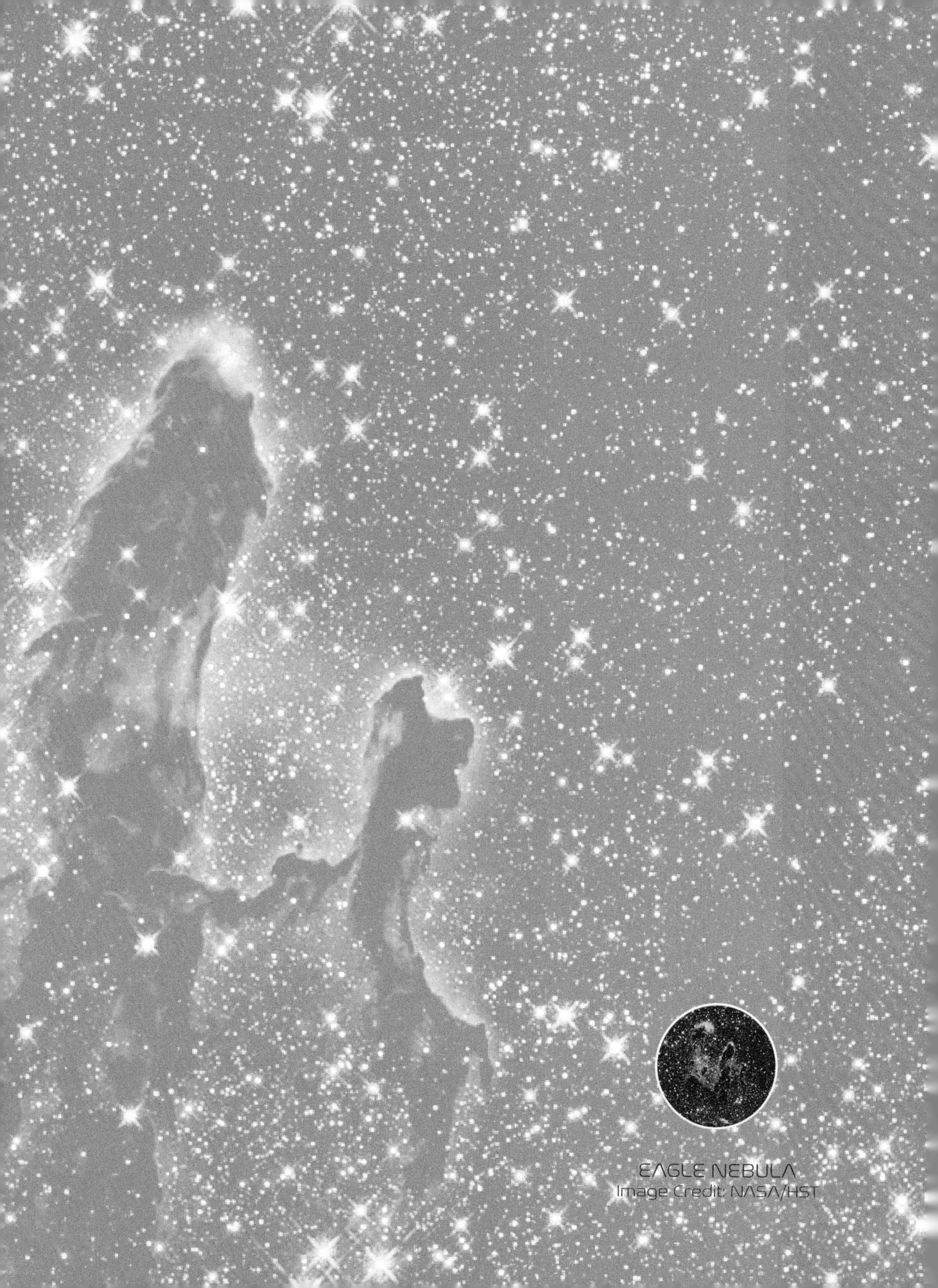

EAGLE NEBULA
Image Credit: NASA/HST

The nitrogen in our DNA, the calcium in our teeth, the iron in our blood, the carbon in our apple pies were made in the interiors of collapsing stars.
We are made of starstuff.
-- Carl Sagan

PLEIADES M45
Star Cluster
Image Credit: NASA/HST

One of the great things about being an astronomer is that you use the telescope to look back in time.
-- Sandra Faber

HUBBLE DEEP FIELD
Image Credit: NASA /
Hubble Space Telescope

Looking at these stars suddenly dwarfed my own troubles
and all the gravities of terrestrial life.
-- H.G. Wells, The Time Machine

The story of humans is the story of ideas that shine light into dark corners ...
We are the products of a billion-year lineage of wandering stardust.
We, all of us, are what happens when a primordial mixture of hydrogen
and helium evolves for so long that it begins to ask where it came from.
-- Jill Tarter

Light is the only connection we have with the Universe beyond our solar system, and the only connection our ancestors had with anything beyond Earth. Follow the light and we can journey from the confines of our planet to other worlds that orbit the Sun without ever dreaming of spacecraft. To look up is to look back in time, because the ancient beams of light are messengers from the Universe's distant past.

-- Brian Cox

We are an impossibility in an impossible Universe.
-- Ray Bradbury

Two possibilities exist;
either we are alone in the Universe, or we are not.
Both are equally terrifying.
-- Arthur C. Clarke

Studying whether there's life on Mars or studying how the universe began, there's something magical about pushing back the frontiers of knowledge.

-- Astronaut Sally Ride

Look at the sky. We are not alone. The whole universe is friendly to us
and conspires only to give the best to those who dream and work.
-- A. P. J. Abdul Kalam

When we contemplate the whole globe as one great dewdrop,
striped and dotted with continents and islands,
flying through space with other stars all singing and shining together as one,
the whole universe appears as an infinite storm of beauty.
-- John Muir

Follow your bliss and the Universe
will open doors where there were only walls.
-- Joseph Campbell

This morning I was the leader of the Universe as I know it.
This afternoon, I'm only a voice in a chorus.
But I think it was a good day.
-- Durken, "First Contact" Star Trek: The Next Generation
{Upon learning of life beyond his home planet.}

When you look at the stars and the galaxy,
you feel that you are not just from any particular piece of land,
but from the solar system.
-- Astronaut Kalpana "K.C." Chawla

LUNAR SURFACE
Image Credit: NASA

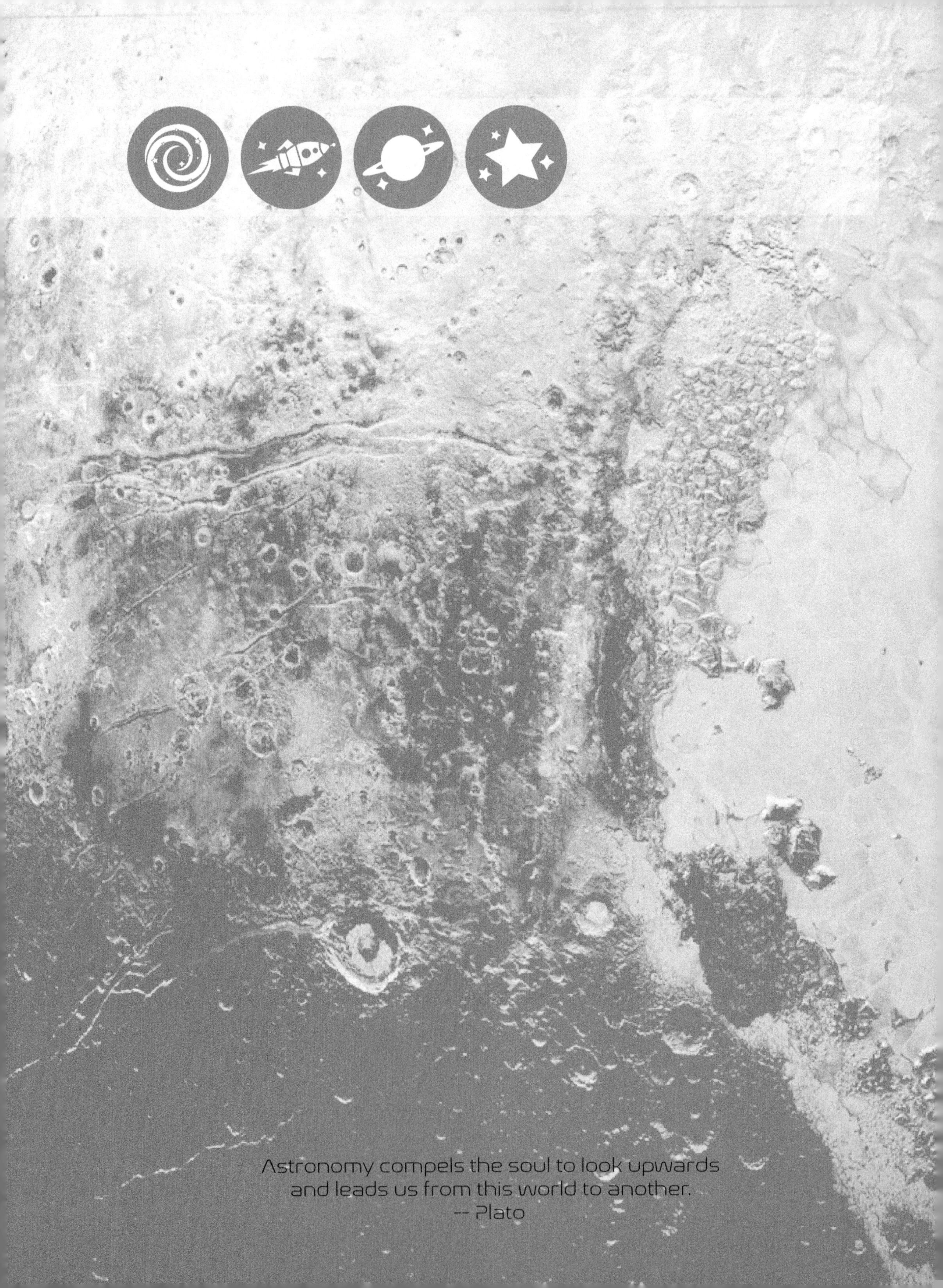

Astronomy compels the soul to look upwards
and leads us from this world to another.
-- Plato

PLUTO's HEART
Image Credit: NASA / New Horizons

People will visit Mars, they will settle mars, and we should because it's cool.

-- Jeff Bezos

MARS SAND DUNES
Mawrth Vallis
Image Credit: NASA/HiRise

Stars are magical. Their light beams out stories and secrets
of ancient times and distant places in our galaxy and beyond.
Look into the night sky for the first star to twinkle 'hello' to you.
Wink back at your new star friend!
And then, get cozy, be patient, and watch and listen for the
secrets your star wants to tell you. What does it say?
Write or draw about it. Create your own magical star tale. "
-- Laura S. Woodmansee

Second star to the right and straight on 'til morning.
-- J.M. Barrie, Peter Pan

NGC 6946
Spiral Galaxy
Image Credit: NASA/HST

Other intelligent lifeforms will differ greatly in appearance.
They may resemble the creature E.T. or startle us with their beauty.
But life itself is common, I'm certain.
-- Frank Drake

SMALL MAGELLANIC CLOUD
Image Credit: NASA/HST

We are all of us stars, and we deserve to twinkle.
-- Marilyn Monroe

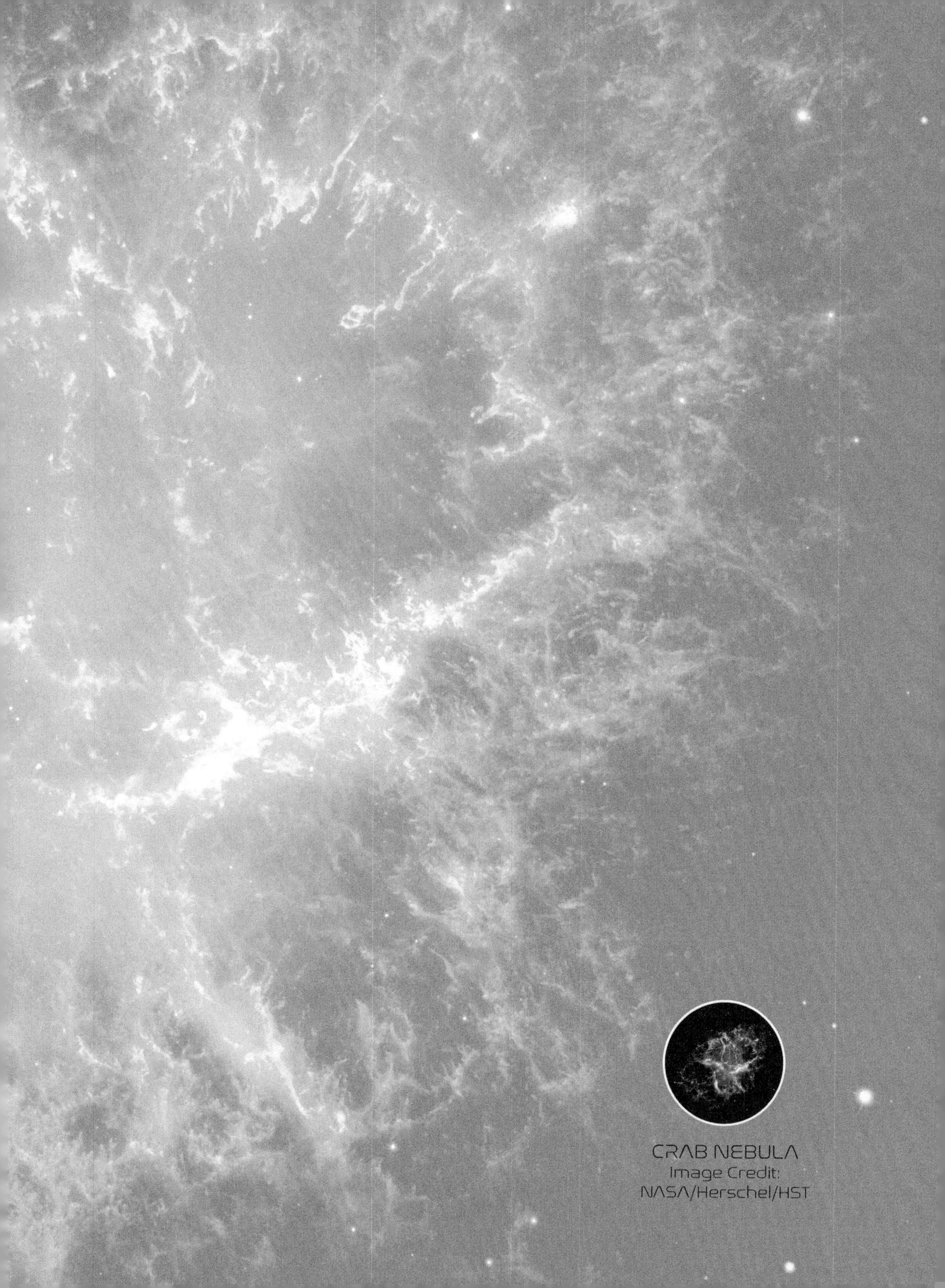

CRAB NEBULA
Image Credit:
NASA/Herschel/HST

We came all this way to explore the Moon,
and the most important thing is that we discovered the Earth.
-- Apollo Astronaut William Anders

WESTERN GLOBE
EARTH
Image Credit: NASA

We are the universe made manifest,
trying to figure itself out.
-- Delenn, Babylon 5

LOBSTER NEBULA
NGC 6357
Image Credit: NASA/HST

We are part of this universe; we are in this universe. But perhaps more
important than both of those facts, is that the universe is in us.
-- Neil deGrasse Tyson

HORSEHEAD NEBULA
Image Credit: NASA /
Hubble Space Telescope

For my part I know nothing with any certainty,
but the sight of the stars makes me dream.
-- Vincent Van Gogh

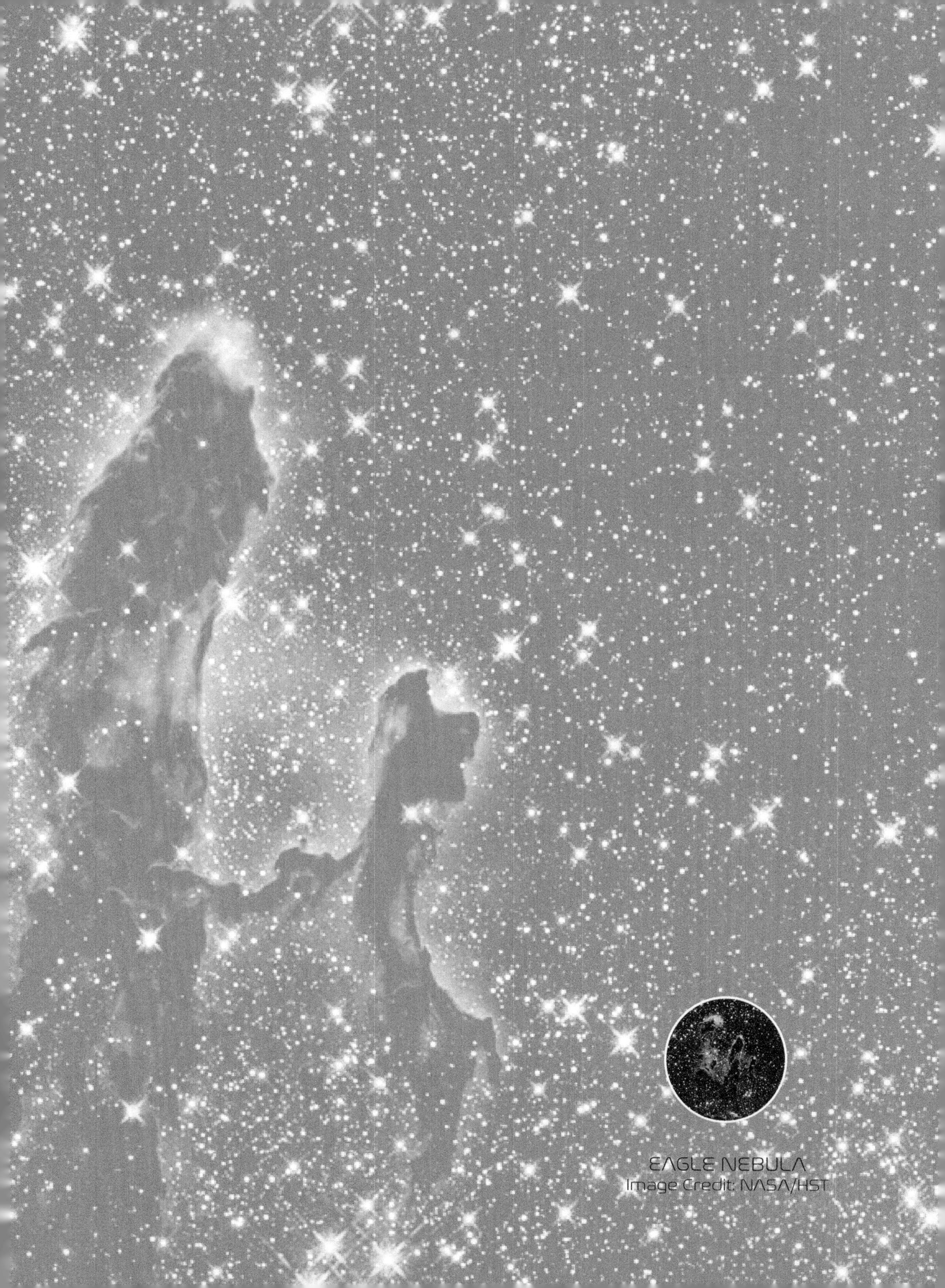

EAGLE NEBULA
Image Credit: NASA/HST

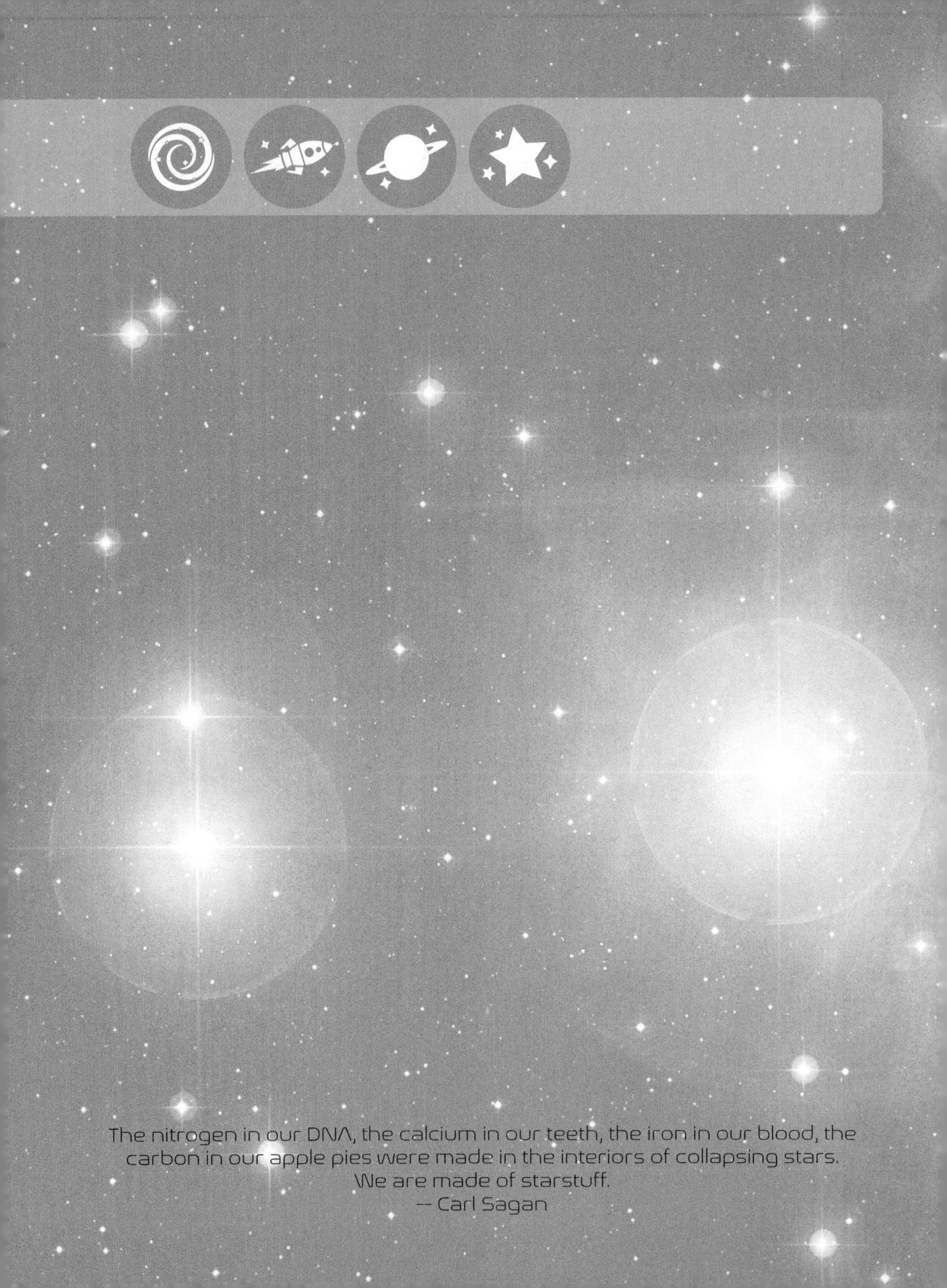

The nitrogen in our DNA, the calcium in our teeth, the iron in our blood, the carbon in our apple pies were made in the interiors of collapsing stars. We are made of starstuff.
-- Carl Sagan

PLEIADES M45
Star Cluster
Image Credit: NASA/HST

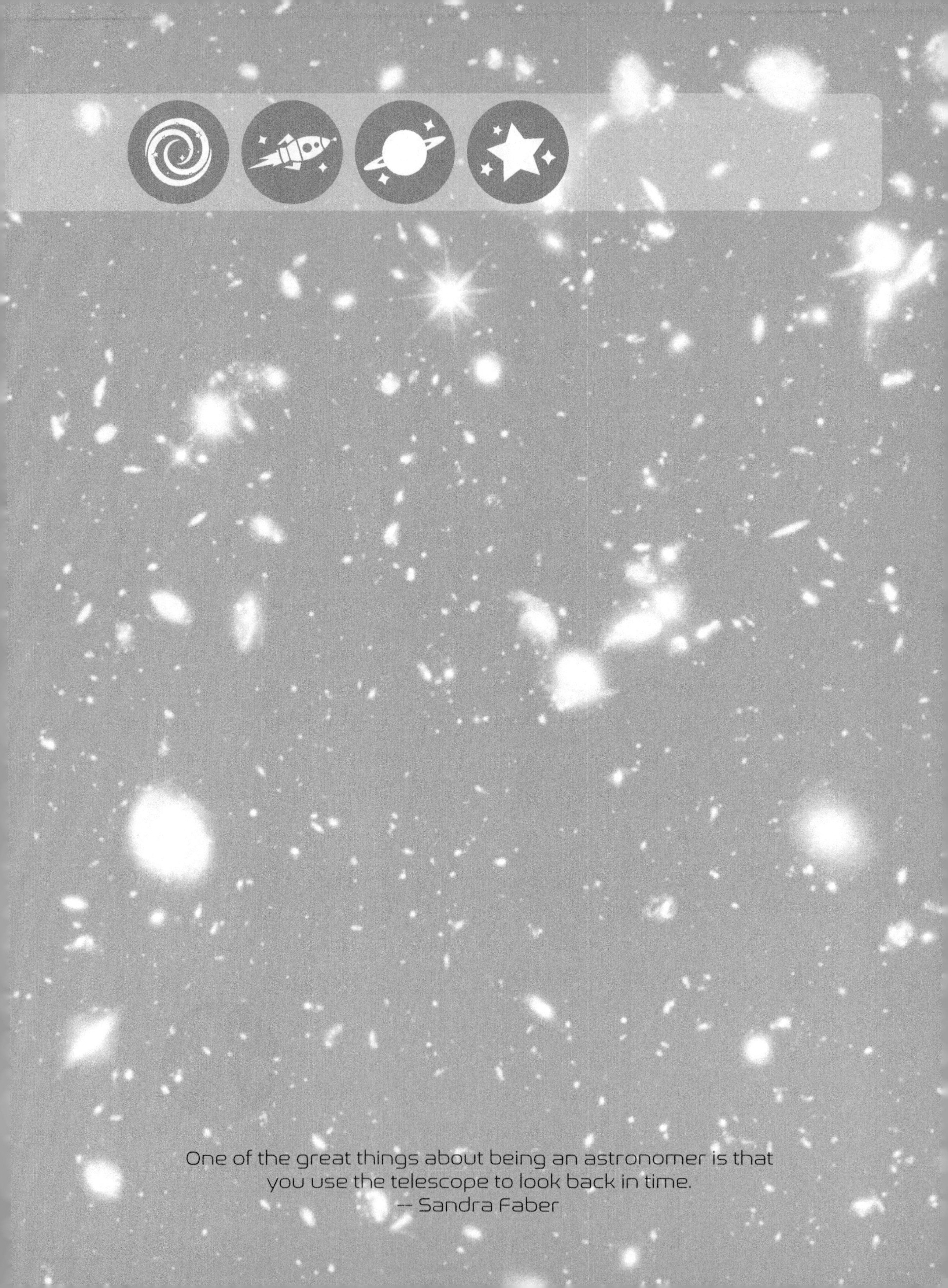

One of the great things about being an astronomer is that you use the telescope to look back in time.
-- Sandra Faber

HUBBLE DEEP FIELD
Image Credit: NASA /
Hubble Space Telescope

Looking at these stars suddenly dwarfed my own troubles
and all the gravities of terrestrial life.
-- H.G. Wells, The Time Machine

The story of humans is the story of ideas that shine light into dark corners ...
We are the products of a billion-year lineage of wandering stardust.
We, all of us, are what happens when a primordial mixture of hydrogen
and helium evolves for so long that it begins to ask where it came from.
-- Jill Tarter

Light is the only connection we have with the Universe beyond our solar system, and the only connection our ancestors had with anything beyond Earth. Follow the light and we can journey from the confines of our planet to other worlds that orbit the Sun without ever dreaming of spacecraft. To look up is to look back in time, because the ancient beams of light are messengers from the Universe's distant past.

-- Brian Cox

We are an impossibility in an impossible Universe.
-- Ray Bradbury

Two possibilities exist;
either we are alone in the Universe, or we are not.
Both are equally terrifying.
-- Arthur C. Clarke

Studying whether there's life on Mars or studying how the universe began, there's something magical about pushing back the frontiers of knowledge.
-- Astronaut Sally Ride

Look at the sky. We are not alone. The whole universe is friendly to us and conspires only to give the best to those who dream and work.
-- A. P. J. Abdul Kalam

When we contemplate the whole globe as one great dewdrop,
striped and dotted with continents and islands,
flying through space with other stars all singing and shining together as one,
the whole universe appears as an infinite storm of beauty.
-- John Muir

Follow your bliss and the Universe
will open doors where there were only walls.
-- Joseph Campbell

This morning I was the leader of the Universe as I know it.
This afternoon, I'm only a voice in a chorus.
But I think it was a good day.
-- Durken, "First Contact" Star Trek: The Next Generation
{Upon learning of life beyond his home planet.}

When you look at the stars and the galaxy,
you feel that you are not just from any particular piece of land,
but from the solar system.
-- Astronaut Kalpana "K.C." Chawla

LUNAR SURFACE
Image Credit: NASA

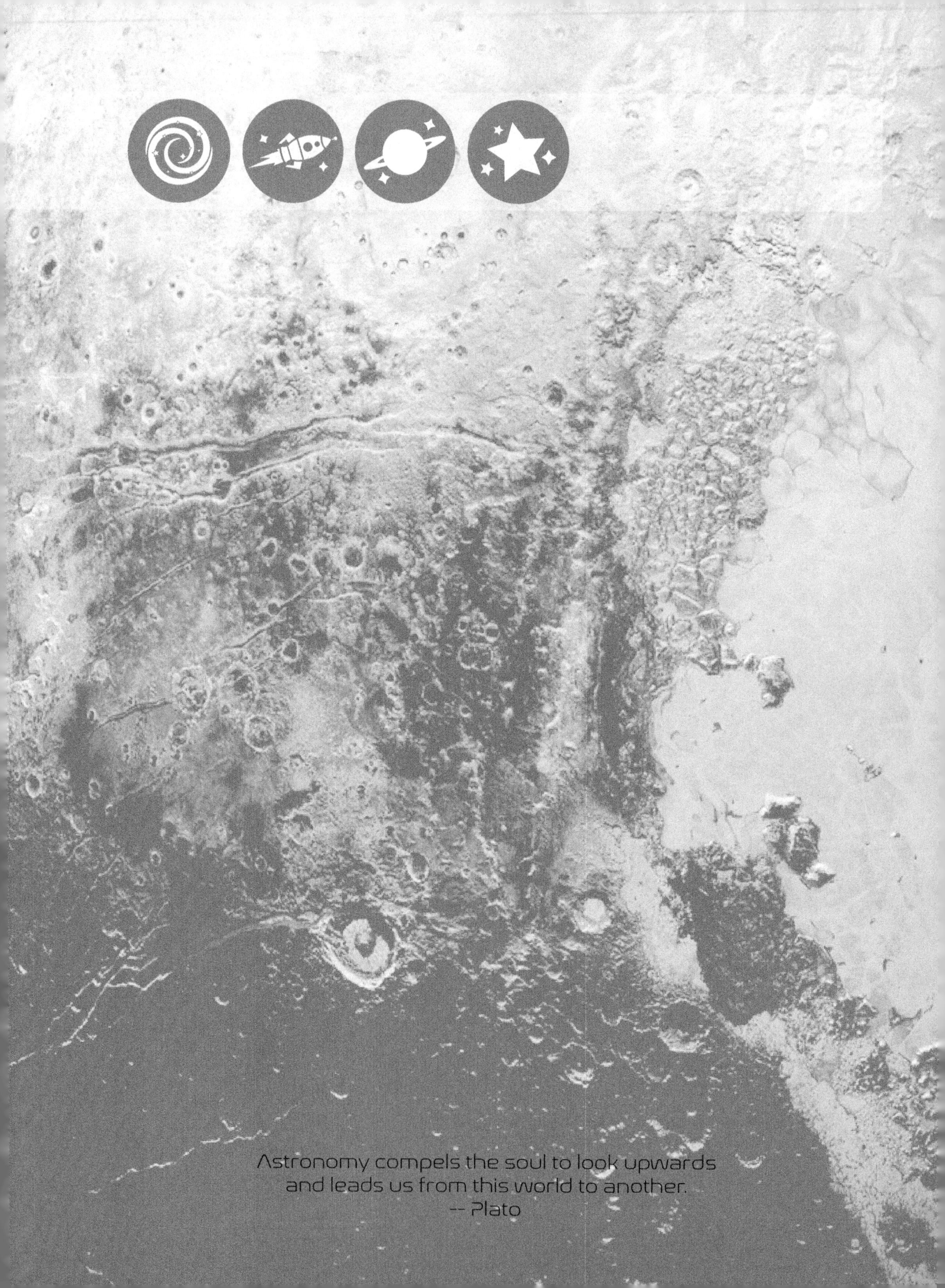

Astronomy compels the soul to look upwards
and leads us from this world to another.
-- Plato

PLUTO's HEART
Image Credit:NASA / New Horizons

People will visit Mars, they will settle mars, and we should because it's cool.
-- Jeff Bezos

MARS SAND DUNES
Mawrth Vallis
Image Credit: NASA/HiRise

Stars are magical. Their light beams out stories and secrets of ancient times and distant places in our galaxy and beyond. Look into the night sky for the first star to twinkle 'hello' to you. Wink back at your new star friend!
And then, get cozy, be patient, and watch and listen for the secrets your star wants to tell you. What does it say? Write or draw about it. Create your own magical star tale. "
-- Laura S. Woodmansee

Second star to the right and straight on 'til morning.
-- J.M. Barrie, Peter Pan

NGC 6946
Spiral Galaxy
Image Credit: NASA/HST

Other intelligent lifeforms will differ greatly in appearance.
They may resemble the creature E.T. or startle us with their beauty.
But life itself is common, I'm certain.
-- Frank Drake

SMALL MAGELLANIC CLOUD
Image Credit: NASA/HST

We are all of us stars, and we deserve to twinkle.
-- Marilyn Monroe

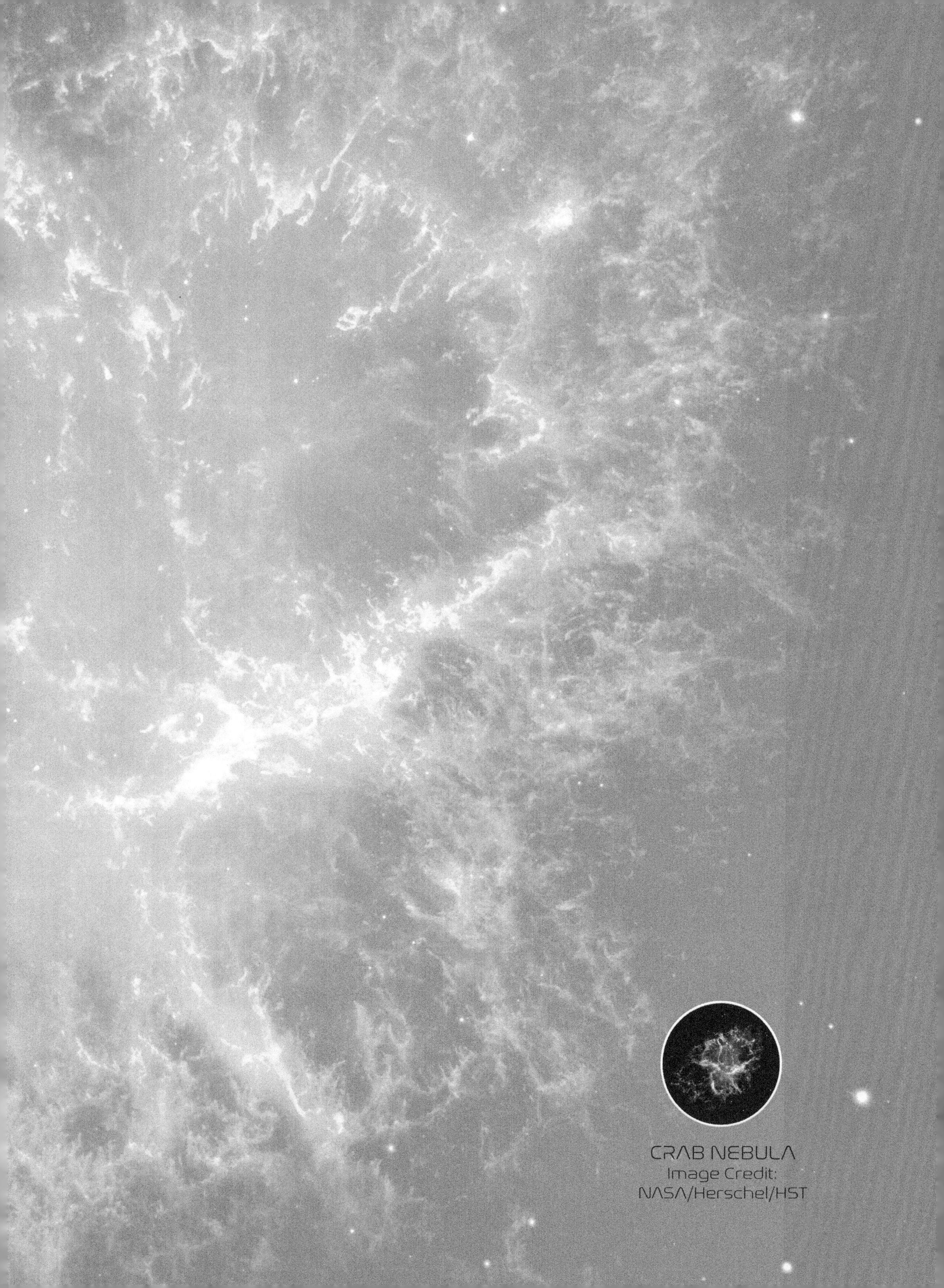

CRAB NEBULA
Image Credit:
NASA/Herschel/HST

We came all this way to explore the Moon,
and the most important thing is that we discovered the Earth.
-- Apollo Astronaut William Anders

WESTERN GLOBE
EARTH
Image Credit: NASA

We are the universe made manifest,
trying to figure itself out.
-- Delenn, Babylon 5

LOBSTER NEBULA
NGC 6357
Image Credit: NASA/HST

We are part of this universe; we are in this universe. But perhaps more important than both of those facts, is that the universe is in us.
-- Neil deGrasse Tyson

HORSEHEAD NEBULA
Image Credit: NASA /
Hubble Space Telescope

For my part I know nothing with any certainty,
but the sight of the stars makes me dream.
-- Vincent Van Gogh

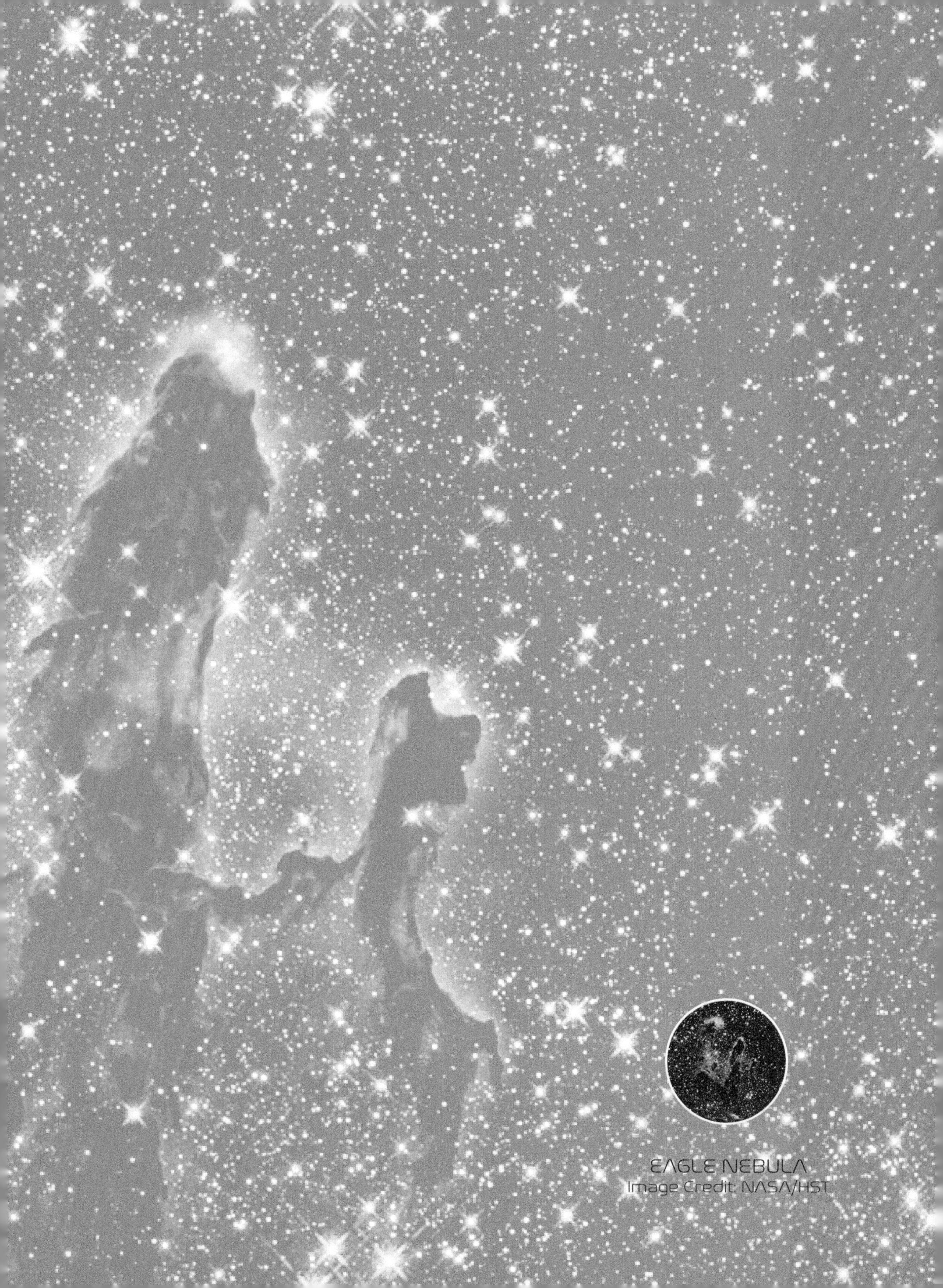

EAGLE NEBULA
Image Credit: NASA/HST

THANK YOU! I hope that you are enjoying this starry little notebook and that the images and quotes fill you with geeky happiness. I had a magical time creating it for you. I listened to lots of cosmic music (check our blog for my playlist) and had a blast researching space images and awesome quotes. Thank you so much to those who I've quoted in this book; several of whom I've been lucky to have as Astronomy teachers back in college and/or as book interviewees over the years as a science writer. Learning from you has added infinite starlight to my life :) And thank you to my husband and my kids; you are awesome and I love that you share in my infinite obsession with space/science/diy/future/sci-fi. Love, Laura

Made in the USA
Monee, IL
08 November 2026

67382869R00063